To my niece Valeria.
For all the happiness she has brought with her.

Latin-Tech Inc
WWW.LT-AUTOMATION.COM

ISBN: 978-0-615935-54-6
PH: 305 848 3517 USA

Robot Story

Miami, FL, USA

Table of Contents

Famous Robots

The following Robots are famous because they were developed by big industrial manufacturers or by research labs, in industries or universities.

We have tried to give you the name of the first one (person or company) who developed the concept or name of the Robot.

In some other cases, when the Robot is designed in a company by an artist, engineer or industrial designer, although the company owns the **rights**, the last one is considered the Robot´s creator.

Puma
Unimate
1966

This is one of the first commercial Robot arms with 6 **degrees of freedom**. It´s used on industrial applications such as:

organizing, moving, loading and assembly of machinery parts.

His inventor, George Charles Devol, Jr, holds the first industrial Robot **patent**.

The company Unimate developed this Robot for General Motors.

Asimo

Honda

1986

Asimo is one of the most innovative Androids in Robotics.

This technological experiment started in1986, and brought a final biped Robot which can walk, run, and climb.

It has face recognition and basic voice commands.

The latest models have less weight (54Kg) and can work for one hour.

It has hands with five fingers which allow him to grasp different objects. Asimo is only 52 inches tall.

Mars Pathfinder

NASA

1992

This Robot was developed by NASA as a **demonstration** of how to land a **rover**, with **sophisticated** instruments and autonomous Robotic functions, on the surface of planet Mars.

It has six wheels and a solar panel.

Aibo
Sony
1999

AIBO is a Robot dog manufactured by SONY Corporation, as a low cost platform for **Artificial Intelligence**.

It was designed for Robots to self-learn and expand their own knowledge.

This autonomous Robot pet is no longer in the market.

Robot **enthusiasts** can download the **programming** tools from SONY's website.

Robonauts
Nasa and GM
2000

The National Aeronautics and Space Administration (NASA), and General Motors (GM) from United States, are developing super advanced Robots to help Astronauts do their jobs in space.

These Robots are called **Robonauts**.

Future space missions will use Robonauts as part of their crew.

This page has been
left blank intentionally.

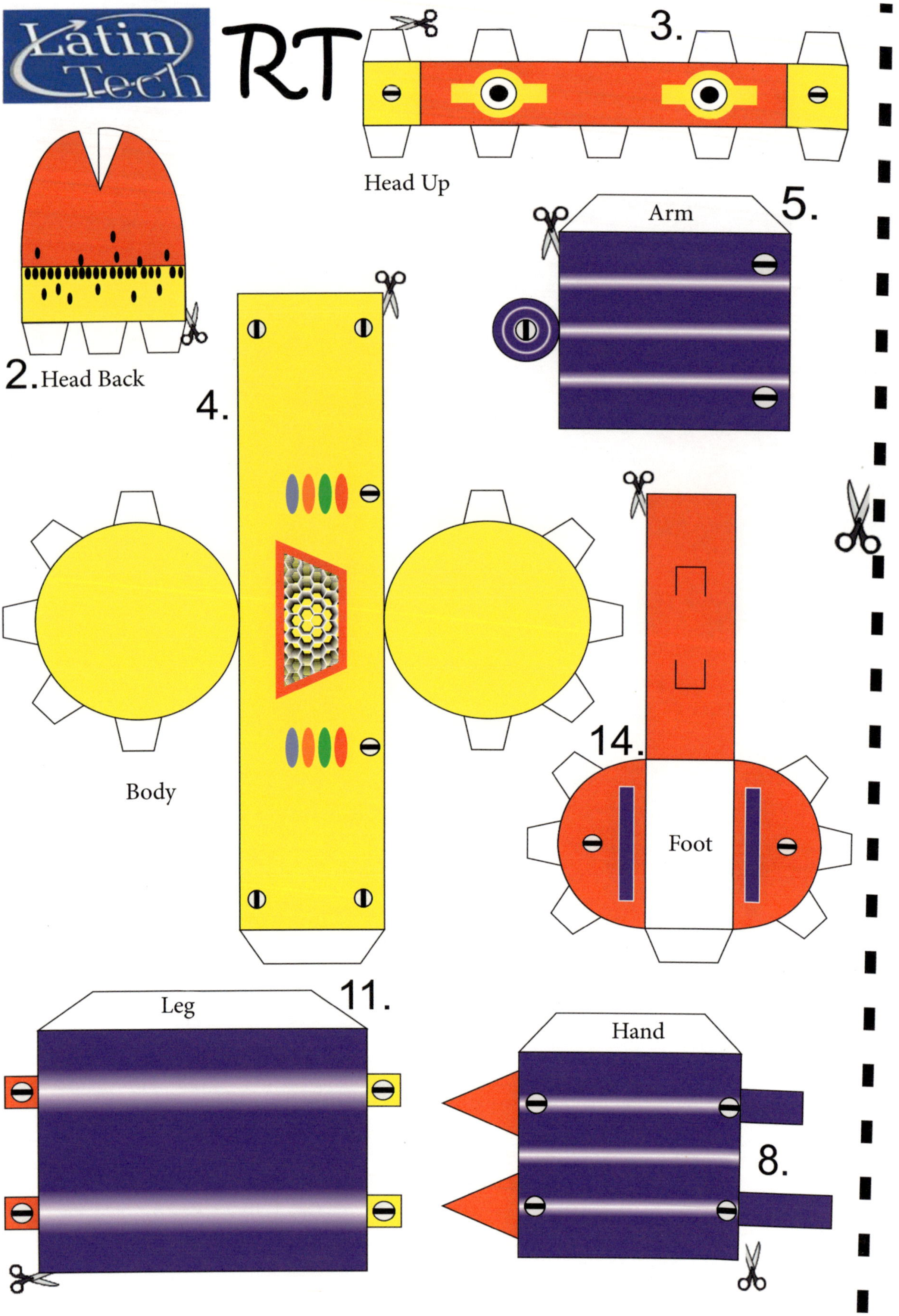

Latin Tech
RT
3.
Head Up
2. Head Back
5.
Arm
4.
Body
14.
Foot
11.
Leg
Hand
8.

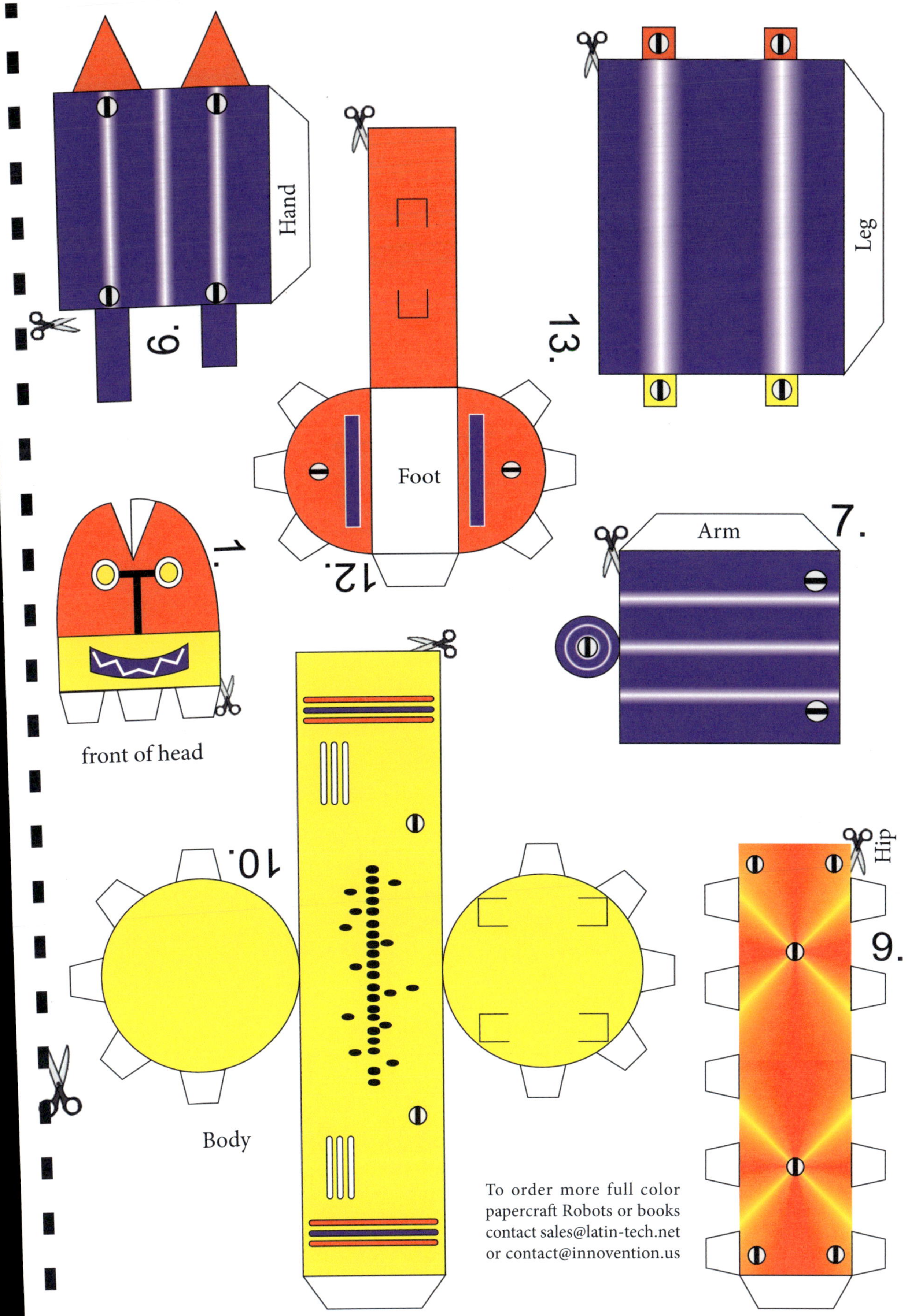

To order more full color papercraft Robots or books contact sales@latin-tech.net or contact@innovention.us

This page has been
left blank intentionally.

Qrio

Sony

2003

This Robot was one of the first models in walking, but it was never launched to the market.

It was manufactured by SONY Corporation, but its production was discontinued in 2006.

QRIO is an **autonomous** android which can recognize voices and faces. It works with batteries during one hour.

Its total height is around 2 feet.

This Robot was, by far, one of the best ways to **visualize** the future… now.

Insbot
IT Haifa
2009

Professor Moshe Shaham, head of the Robotics Laboratory in Israel, and his team from the Robotics laboratory at Institute of Technology in Haifa Technion, have developed a very small Robotic fly which can be introduced in your body to **unclog arteries** and inject mini **doses** of medicines on **infected** parts.

Measuring only 1 mm, this Robot can fly at a speed of 9mm per second while is controlled by an external **magnetic field**.

This tiny Robot is based on Micro-Electro-Mechanical Systems (MEMS) technology.

RiSe
DARPA
2005

RiSe is a new Robot developed by Boston Dynamics in **collaboration** with University of Pennsylvania, Carnegie Mellon, UC Berkeley, Stanford, and Lewis and Clark University.

It was **funded** by the Defense Advanced Research Projects Agency DARPA.

Boston Dynamics is a company which specializes in advanced Robots with very sophisticated mobility, **agility**, **dexterity** and speed.

It has six legs, each of which is powered by two electric motors. The Robot can adapt its posture according to the surface it´s climbing.

It looks like a gecko lizard, but the real animal only has four legs.

BigDog
DARPA
2005

This Robot is also funded by DARPA and developed by Boston Dynamics. BigDog is a 75 cm tall and 90cm long, four legs, walking Robot.

This Robot is very strong and looks like a small **mule**.

Its weight is 240 pounds.

It can run at 4Mph, climb **slopes** up to 35 degrees, walk in snow and surfaces with water, and carry heavy loads up to 340 lbs on any type of surface.

This Robot has very high technology sensors and controls for its autonomous operation.

Recently, the same company has launched another similar, but smaller size Robot version named “Littledog”.

Mini Robots Army
University of Pennsilvania
2011

Alex Kushleyev and Daniel Mellinger from the General Robotics, Automation, Sensing and **Perception** (GRASP) Laboratory at the University of Pennsylvania, recently introduced very tiny, but **smart**, flying mini Robots.

They have sensors and complex controls to be able to fly, **rotate** and travel through **obstacles**. They can also cooperate with other Robots as a team.

They can be used in difficult situations when human life could be in danger, like the case of fire or **radioactivity**.

They also can do creative things, like forming an **orchestra** to play different musical instruments.

Flying Micro Robots

Waterloo University

2009

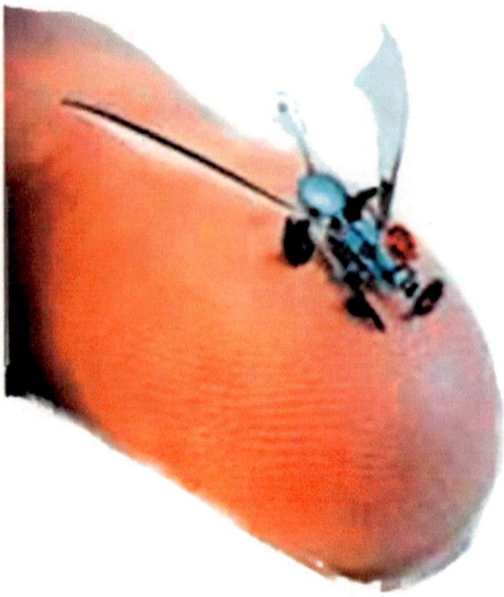

Canadian researcher Behrad Khamesee and his team, from Waterloo University, have developed a flying micro-Robot which can **manipulate** objects and can carry very small tools which can be used on many micro scale applications like **surgeries**, assembly of tiny mechanical parts or helping in human body diagnostics.

This micro Robot is based on Micro-Electro-Mechanical Systems (MEMS) technology as well.

The Robot is powered by a magnetic field to fly.

The Robot has sensors and very sophisticated electronics.

HRP-4C

AIST

2009

The National Institute of Advanced Industrial Science and Technology (AIST) in Tokyo, Japan has developed a Robot which really looks and moves like a person.

It´s an android girl named HRP-4C.

The Robot weighs 95 lbs and is 5'2 tall.

The wonder of this Robot is that she can show some **emotions** or expressions on her face.

When she walks, it looks like a real human is walking.

She can also sing and dance.

Runner Robots

A very particular Robotic contest took place in Osaka, Japan:

The world's first full **marathon** for Robots.

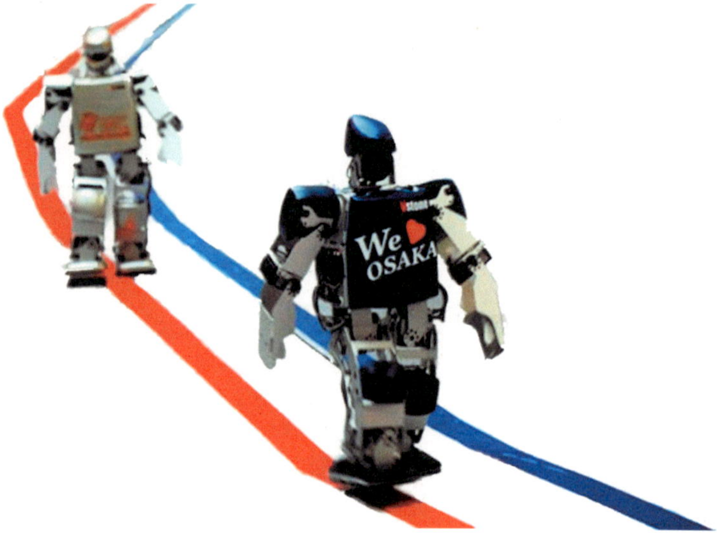

Several biped Robots, of the height of a knee, have to run a marathon of 422 laps on an indoor track.

The Robot's owners are only allowed to replace the batteries or to change the motors. If for any reason, the Robot falls over, it must be able to get up by itself otherwise gets out of the competition.

Some Robots practiced and **stretched** their legs before joining the competition.

Vocabulary

Agility: Capacity to make easy and quick moves.

Arteries: Conducts that carry blood away from the heart.

Artificial: Something made by humans.

Autonomous: The Robot that can make decisions by itself.

Biped: With two legs.

Collaboration: To work together.

Commands: An order given with authority.

Cooperate: To work together.

Crew: Group of persons that control a vehicle or a ship.

Enthusiasts: People who like something very much.

Expand: To have more or wider knowledge.

Degrees of freedom: It is the number of possible independent movements the Robot arm can do.

Demonstration: Prototype, Unit for tests.

Dexterity: Movements with precision.

Diagnostics: Identification of an illness.

Download: To get software from the web.

Experiment: Scientific procedure to check or discover something.

Face recognition: When a Robot is able to know the person before of him.

Grasp: To hold firmly with the hand.

Humanoid: That looks like a human being.

Innovative: Something that is new and that people like.

Indoor track: A location with a high roof where a competition takes place.

Infected: Contaminated with virus or microrganisms.

Laps: One complete circle in a race.

Magnetic field: An invisible force similar to the one produced by magnets.

Manipulate: to take and move to a different place.

Marathon: A very long distance race.

Micro scale: A much smaller size.

Mini doses: Tiny portions of medicine.

Mule: The animal that comes from a female horse and a male donkey.

Obstacles: Things that block the pass.

Orchestra: Group of musicians.

Patent: Document that tells that a person was the first in the idea or its implementation.

Perception: The idea you get from something using your senses.

Platform: Basic system to help in the development of more Robots.

Vocabulary

Posture: Position of the body or body parts.

Powered: That has power to operate.

Programming: Instructions to automate tasks.

Radioactivity: Invisible and dangerous atomic particles that come out from some elements.

Rights: A moral or legal permit to do or to have something

Robonaut: Robot designed by NASA to be used in space missions.

Rotate: To turn around a center.

Rover: Vehicle with wheels to move on difficult surfaces.

Self-learn: To get the knowledge by oneself.

Sensors: Devices that help to detect what is around you.

Sophisticated: Something of high complexity.

Slope: Inclined surface.

Smart: Pretty intelligent Solar.

panel: A device which converts sunlight into electricity.

Strech: Extend one´s body to its full length.

Surgeries: Operation or procedures to remove a part or a total of a given body part or organ.

Unclog: To remove something that is blocking the pass.

Visualize: To imagine.

Voice commands: Giving orders to Robots using voice.

Website: Place on the internet about a company or a product.

RT

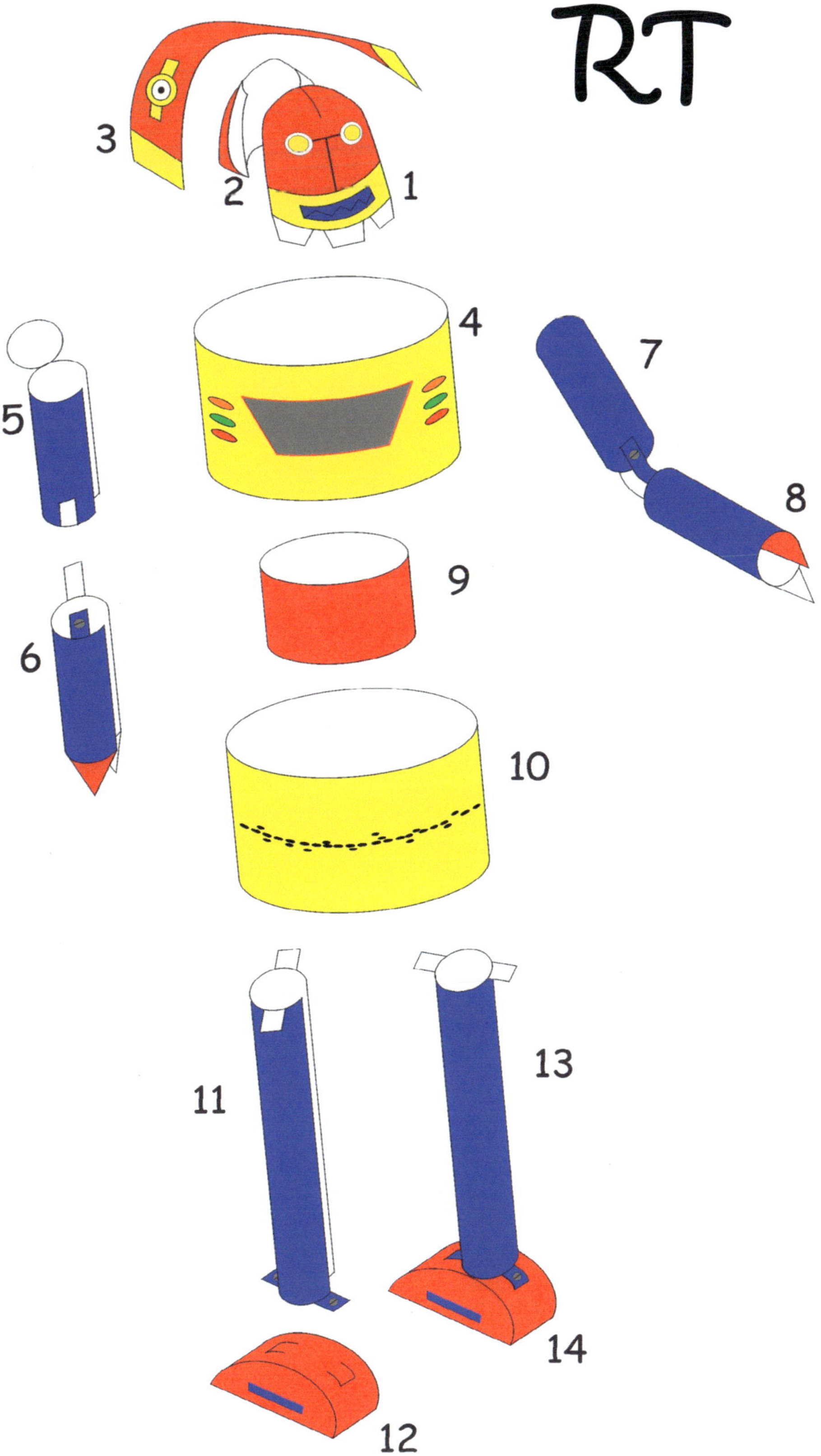

Made in the USA
Monee, IL
05 November 2026

67382738R00017